AF270666

LOCH NESS MONSTER

SCOTLAND'S SEA SERPENT

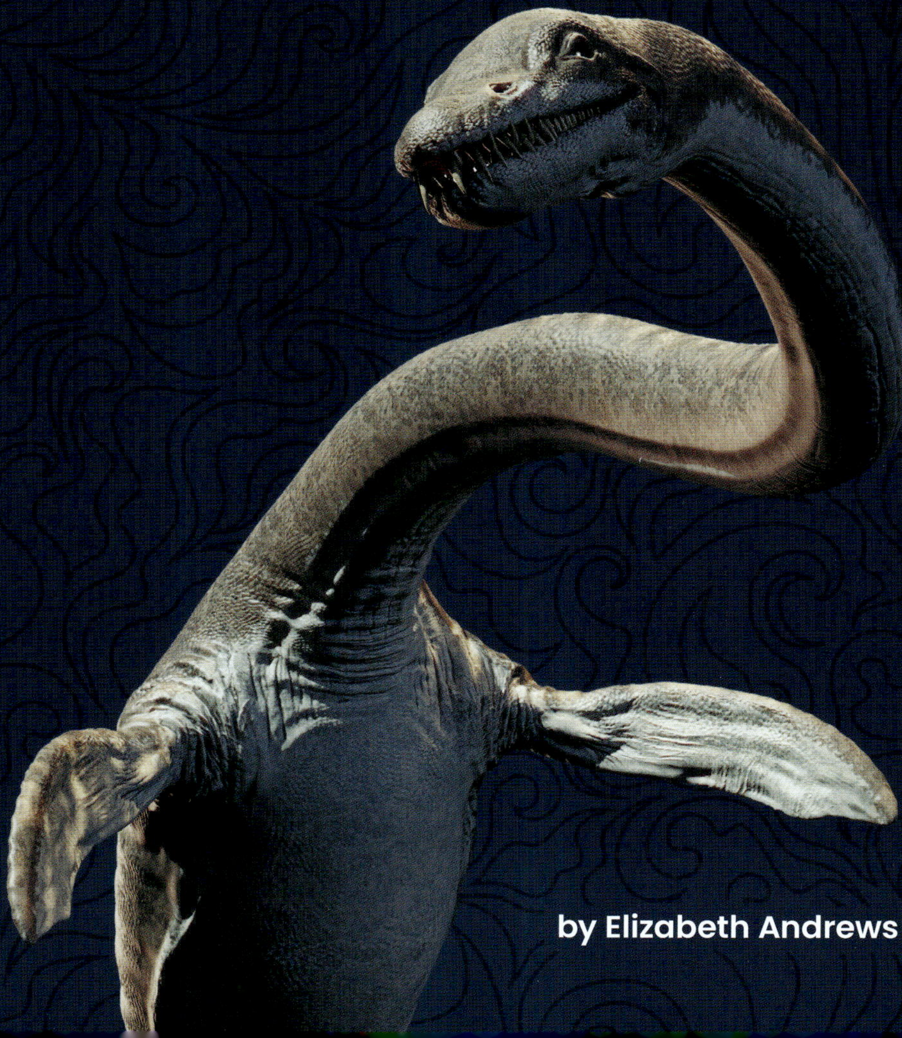

by Elizabeth Andrews

abdobooks.com

Published by Pop!, a division of ABDO, PO Box 398166, Minneapolis, Minnesota 55439. Copyright © 2023 by Abdo Consulting Group, Inc. International copyrights reserved in all countries. No part of this book may be reproduced in any form without written permission from the publisher. DiscoverRoo™ is a trademark and logo of Pop!.

Printed in the United States of America, North Mankato, Minnesota.

052022
092022

THIS BOOK CONTAINS RECYCLED MATERIALS

Cover Photo: Shutterstock Images

Interior Photos: Shutterstock Images, Getty Images, Biodiversity Heritage Library, Edwin Sampson/ANL/Shutterstock, Historia/Shutterstock, Daily Mail/Shutterstock, James Gray/Shutterstock, John Dee/Shutterstock

Editor: Bridget O'Brien
Series Designer: Laura Graphenteen

Library of Congress Control Number: 2021951854

Publisher's Cataloging-in-Publication Data

Names: Andrews, Elizabeth, author.

Title: Loch Ness Monster: Scotland's sea serpent / by Elizabeth Andrews

Other title: Scotland's sea serpent

Description: Minneapolis, Minnesota : Pop, 2023 | Series: Creatures of legend | Includes online resources and index

Identifiers: ISBN 9781098242367 (lib. bdg.) | ISBN 9781098243067 (ebook)

Subjects: LCSH: Loch Ness monster--Juvenile literature. | Nessie (Monster)--Juvenile literature. | Sea serpents--Juvenile literature. | Monsters--Scotland--Juvenile literature. | Animals--Folklore--Juvenile literature. | Fabled creatures--Juvenile literature.

Classification: DDC 001.944--dc23

WELCOME TO DiscoverRoo!

Pop open this book and you'll find QR codes loaded with information, so you can learn even more!

Scan this code* and others like it while you read, or visit the website below to make this book pop!

popbooksonline.com/loch-ness

*Scanning QR codes requires a web-enabled smart device with a QR code reader app and a camera.

TABLE OF CONTENTS

ATTRACTION

It was a classic foggy morning in the Scottish **Highlands**. Clouds hung low against the backdrop of emerald-green mountains just outside the town of Inverness. The air held on to the chill of night as small rays of sunlight cracked

through the haze atop Loch Ness. Along

the shores of the lake, people were

setting up deck chairs.

In a single day around Loch Ness, people can experience sun, rain, and snow!

This had been the town's routine for months. After the new road around Loch Ness had been laid, people began arriving from everywhere. They all hoped to catch sight of the **elusive** Nessie.

DID YOU KNOW? People **affectionately** refer to the Loch Ness Monster as Nessie.

People sipped warm tea. Newspaper photographers kept their cameras ready just in case today would finally be the day. The day the Loch Ness Monster broke the surface with her long neck or large fin. But like every other morning, she remained hidden. The mystery of whether there is truly a massive **aquatic** beast swimming beneath the waters would live on another day.

People used the hills and cliffs around Loch Ness to get good lookout points.

NESSIE'S HISTORY

The Loch Ness Monster mystery began 1,500 years ago. People of the British-Roman Empire explored the land that is now Scotland. In 500 CE, the first record of a large **aquatic** creature in Loch Ness was discovered on a stone carving. The carving came from

a group of people living in the **highlands** of Scotland called the Picts.

PAINTED PEOPLE

Picts were farmers who lived in small communities in eastern Scotland during the Dark Ages. They stand out in history because of their beautiful stone artwork and the art inked upon their skin. Their name is likely from the Latin word *picta* which means "painted."

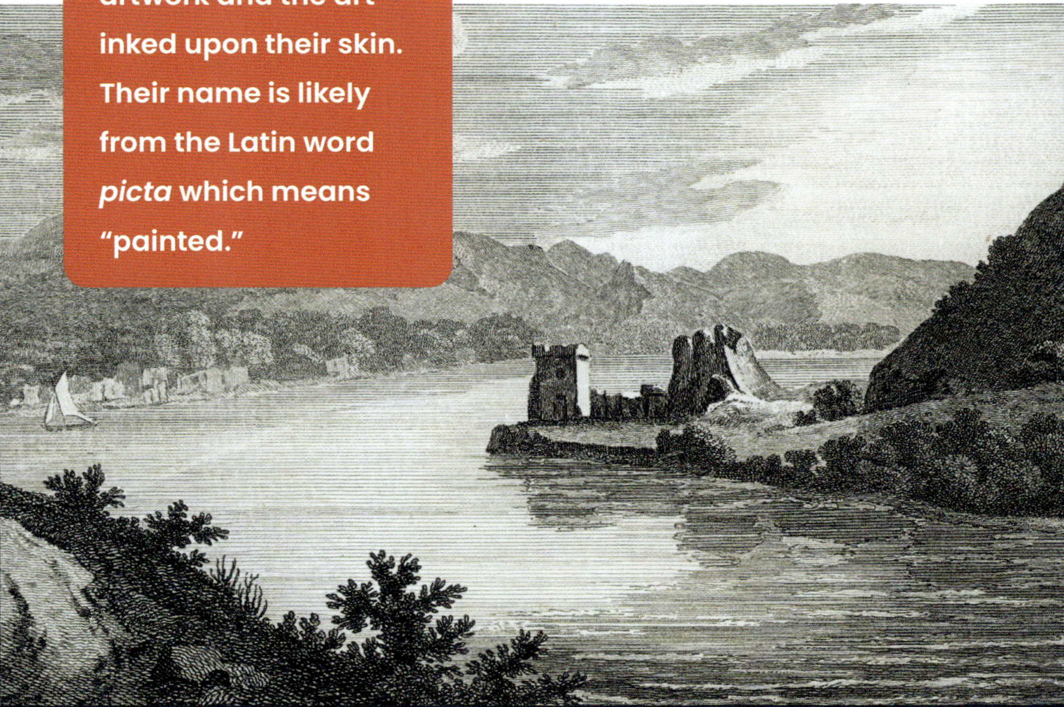

Urquhart Castle was built in the 1200s and has survived many battles.

Picts were known for their animal carvings. Most found pieces have easy-to-identify animals from eastern and northeastern Scotland upon them. However, there is one carving of a mysterious animal with a long muzzle, head spout, and flippers instead of feet. Some who have seen the carving describe it as a swimming elephant. There has been no **conclusive** decision on what animal it is.

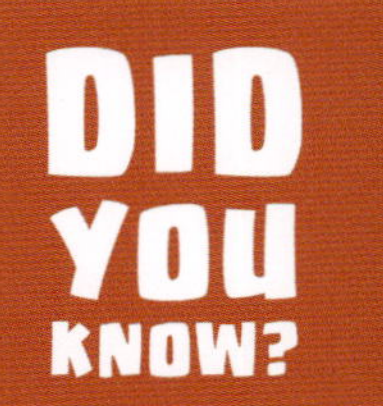

Lake Champlain in Vermont has had kelpie sightings.

Loch Ness is a beautiful home for all creatures that live in it.

The people in Scotland in 500 CE didn't have the advanced science and research skills that exist now. A giant aquatic beast living in Loch Ness wouldn't be too unbelievable. For example, the Scottish passed around tales of fantastic creatures called water kelpies.

Water kelpies are shape-shifting spirits that live in bodies of water. They are dangerous. Kelpies appear as beautiful horses already saddled and inviting people to ride them. Once a person was on the kelpie's back, the spirit would jump into the water and drag the human down to the depths to be eaten. Some people believe that Nessie could be a water kelpie.

Kelpies may
sometimes
appear as
a human.

THE NESSIE CRAZE EXPLODES

In 1933 a road was laid around Loch Ness. More people spent time looking at the water as they drove along the new roadway. The number of Nessie sightings went way up.

There were two exciting accounts during the first year. One was from a local

couple who saw "an enormous animal rolling and plunging in the surf." Another couple saw a large animal that looked like a dragon walking in front of their car along the new road. A newspaper reported these sightings, and the whole country got excited about a beast in Loch Ness.

A82 is the main road people spot Nessie from.

That same year, a famous hunter named Marmaduke Wetherell went searching for the monster. He quickly found footprints of a large beast and made **casts** of the tracks. However, scientists from the Natural History Museum **debunked** the tracks. All four footprints Wetherell found were made from a single stuffed hippopotamus foot. No one knows who made the footprints.

In 1934, the infamous Surgeon's Photograph of Nessie was taken. This image inspired new theories about what kind of creature lived in the lake. The long neck rising out of the water resembled a plesiosaur. People believed Nessie may be a dinosaur that didn't die out 66 million years ago. This photo was trusted by nearly everyone because it came from a reliable source.

This is the Surgeon's Photograph taken by R. Kenneth Wilson.

In 1994 the Surgeon's Photo was proved to be a **hoax** put on by the *Daily Mail* newspaper. Loch Ness Monster believers were not deterred though. New sightings kept rolling in.

CHAPTER 4

WHO IS SHE REALLY?

Before the Surgeon's Photograph was **debunked**, many were invested in uncovering the truth of Nessie. There were so many reports of an enormous **aquatic** reptilian creature that a scientific investigation of Loch Ness was necessary.

Is this proof that the Loch Ness monster DOES exist?

By **Fidelma Cook** and **Ian Gallagher**

FROM the murky waters of Loch Ness, a neck-like shape emerges, standing almost on end, then arching forward and vanishing slowly below the surface.

And last night the question was being asked: Could this be the proof we've been waiting for that the elusive Scots beast really *does* exist?

The debate was sensationally reignited after these astonishing photographs were captured by James Gray, a respected cameraman with 30 years' newspaper experience who took the first pictures of Prince Charles and Diana together when she was a shy 19-year-old girl desperate to make the future King notice her.

Experts on both sides of the fence are already said to be 'excited' by the clarity of the Loch Ness images, which have previously tended to be tantalisingly vague.

But 54-year-old Gray, who never goes anywhere without a camera, was not using state-of-the-art equipment or telephoto lenses . . . merely an old Nikon 801 stuffed into a fishing bag in his little boat.

'It was an extraordinary moment,' he said yesterday. 'I'd never believed in Nessie, but I'm convinced

'This was a lot more exciting than any Royal exclusive'

that these pictures are more important than any Royal scoop.'

Gray, a keen salmon fisherman who lives in semi-retirement at Invermoriston village, 20 miles from Inverness, had often taken his 16ft boat to the point where the River Moriston flows into Loch Ness.

Earlier this month he and local barman Peter Levings ended up about three quarters of a mile from the river join.

'We usually go out at the same time, around 6am when it's calm and before the wind gets up,' he said. 'That day the conditions were absolutely peaceful. We stopped the motor and trawled around with a fishing lure. Peter was rowing with his back to the front of the boat and I was facing him when we arrived in the middle of the loch.

'Then, suddenly, I spotted a bizarre movement 150 yards away and saw this thing sticking out. I thought it was a duck or something. I instinctively grabbed my camera and started snapping. The thing – whatever it was – raised itself a couple of feet and was rising even more as I looked at it. Soon, it was about 6ft out of the water.

'I'd stood over Peter to do the pictures. As he turned round to see what I was photographing, he just saw what was left of it . . . a sort of black blob as it disappeared. It had curled forward and gone down.'

Gray added: 'This was certainly no seal – it was too high up in the water for one thing. It had a long black neck almost like a conger eel, which would be impossible in Loch Ness. But I couldn't see a head. It didn't seem to bend very much but as it went under it sort of arched and disappeared.'

'We started the engine and drove to the spot where it had vanished,

'Soon it was rising 6ft out of the water'

circling for around 20 minutes, but we found nothing.'

Gray didn't bother to develop the film immediately because he was sure the image would be too small to have any impact.

'But when I saw the images I got a bit more excited,' he said.

Gray said that for years he had barely glanced at the Nessie-hunters who spend hours on the lochside armed with enormous lenses in their attempts to photograph the enduring mystery.

For a man so used to following celebrities such as Diana there was little lure in monster-hunting.

'I've got better things to do with my time,' he said.

'I always thought that if there is one, then there would have to be more than one – and surely they'd have been seen much more often than is claimed.

'And now? Well, if so many people think it is Nessie, it's a photographer's dream.

'But I'm still not sure that if I've captured the monster on film? I certainly don't know what the damn thing was. Let's just say I'm cautiously excited, but I'm not putting any name to what's in those pictures.'

Gray hasn't spotted the 'creature' since his trip.

But, then again, he's been busy with something rather more important to celebrate just lately – the birth of his daughter.

However, he and his wife, Ulrike, resisted the urge to call the baby Nessie. 'We settled for Simone instead,' he said.

STRANGE ENCOUNTER: James Gray, right, on the loch with Peter Levings

RISING FROM THE DEEP: The figure in the main picture appears to arch up to about 6ft before dropping back down again, right, and vanishing under the water

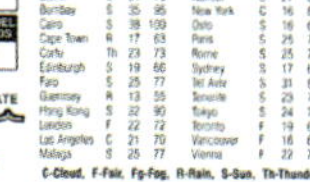

World Weather at noon yesterday						
Athens	S	24	75	Malta	S	23 73
Beijing	S	26	79	Miami	C	28 82
Berlin	S	21	70	Moscow	F	9 48
Bombay	S	35	95	Nairobi	S	27 81
Cairo	S	38	100	New York	C	16 61
Cape Town	R	17	63	Paris	S	25 77
Corfu	Th	23	73	Rome	S	25 77
Edinburgh	S	19	66	Sydney	S	17 63
Faro	S	25	77	Tel Aviv	S	31 88
Guernsey	R	13	55	Tenerife	S	23 73
Hong Kong	S	32	90	Tokyo	S	24 75
London	F	22	72	Toronto	F	19 68
Los Angeles	C	21	70	Vancouver	F	16 61
Malaga	S	25	77	Vienna	F	23 72

C-Cloud, F-Fair, Fg-Fog, R-Rain, S-Sun, Th-Thunder

Roadwatch — from AA Roadwatch

M2 Kent: Jns 1-4 (Rochester-Gillingham)	M42 Warwicks: Jns 10-11 (Tamworth Services Maschani)
M4 Swansea: Jn 45 (Ynysforgan Interchange)	A1 Lincs: Grantham
M5 Somerset: Jns 19-20 (Gordano Services-Clevedon)	A12 Essex: Brentwood
M5 W Mids: Jns 1-3 (West Bromwich-Birmingham)	A13 London: Between Barking and Rainham
M6 Cheshire: Jns 22-24 (Warrington-Ashton)	A26 Antrim: Between Antrim and Ballymena
M6 Lancs: Jn 29 (M65 interchange)	A36 Wilts: Salisbury
M25 Herts: Jns 17-16 (Rickmansworth-M40)	A40 London: Hangar Lane Underpass
	A50 Leics: Leicester
	A189 Northumberland: Cramlington

Still baffling experts after 1,500 years

IF THE extraordinary images prove to be of Nessie, they will end a mystery dating back nearly 1,500 years.

The Mail on Sunday asked five Nessie experts for their opinions on the shots...

● Naturalist Adrian Shine – the world's leading authority on the monster – said: 'It appears too rigid to be a living creature. And there's no head – if anything, it gets thinner at the top. It is probably a tree branch.'

● Rip Hepple, who has spent 35 years hunting Nessie: 'It looks like the bumper off a Morris Minor!'

● Biologist David Martin, who has written a book about the beast: 'It's too stiff. And there seems to be no movement below the water surface. Perhaps it's a log.'

● Gary Campbell of the Official Loch Ness Monster Fan Club: 'What concerns me is the shots do not show the shape disappearing completely. The sequence should show the water after it has vanished. I've no idea what it is.'

● Loch Ness Investigator Dick Raynor: 'I don't think it is Nessie. It could be anything from a tyre to a tree branch.'

Although St Columba is said to have encountered the beast in 565 AD, the modern legend of Loch Ness dates from 1933, when a couple spotted 'an enormous animal rolling and plunging on the surface.'

Their account appeared in the Inverness Courier, which first used the word 'monster'. The most famous picture of Nessie, taken in 1934 by a surgeon, was revealed 60 years later as a fake.

Scientists have consistently dismissed sightings of them as hoaxes or optical illusions caused by floating logs, otters, ducks, or swimming deer.

But based on the strength of sightings in 1993, bookmakers William Hill cut the odds against Nessie being found from 500-1 to 100-1. If Nessie was proven to exist, it faces a payout of more than £1 million.

FAKED: The famous 1934 shot

James Gray and Peter Levings (pictured in boat) had a Nessie sighting while fishing.

Some people believe in a family of Nessies. They don't think an animal could survive all these years, so it must be a species that can reproduce.

In 1987, 20 boats swept the entire Loch from top to bottom using **sonar**. They were looking for signs of large targets deep under the surface of the lake.

During this sonar scan, three unknown targets were found. One was large enough to excite researchers with the possibility it was Nessie. But when they scanned the same area, the target could not be found again.

No photos of Nessie are clear enough to prove she is real.

Another investigation was held in 1990. This time scientists studied the **biology** of Loch Ness. The goal was to learn about all the living things inside the lake. They also studied physical formations and depth changes of the lake. While the scientists were not specifically looking for Nessie, they did come across a large moving creature and followed it for 10 minutes before losing sight. The creature was 15 feet (4.6m) long.

Nessie is often shown as a gentle giant in media.

Despite the sightings, studies, and photographs, the Loch Ness Monster has still not been proven to exist as of 2022. However, the mystery of Nessie has not been officially debunked either. Loch Ness is gigantic and gets as deep as 750 feet (229m) at some points. A mysterious **aquatic** beast could be hiding anywhere.

DID YOU KNOW?

Gary Campbell of Inverness, Scotland, tracks all Nessie reports on his Official Loch Ness Monster Sightings Register.

TEXT-TO-SELF

Do you think the Loch Ness Monster exists? Please explain your answer.

TEXT-TO-TEXT

Have you read any other books about aquatic monsters? If so, what did those monsters have in common with Nessie?

TEXT-TO-WORLD

Why do you think the Loch Ness Monster legend was able to capture the world's attention?

GLOSSARY

affectionately — showing love and care.

aquatic — growing or living in water.

biology — the study of living things.

cast — a reproduction of something by using plaster or metal.

conclusive — serving to reach a final answer or decision.

debunk — to expose falseness.

elusive — hard to find or catch.

highlands — a part or region of a country that has many hills or mountains.

hoax — an act meant to trick.

sonar — a way to find objects underwater by sending and reflecting sound waves.

INDEX

ONLINE RESOURCES

popbooksonline.com

Scan this code* and others like it while you read, or visit the website below to make this book pop!

popbooksonline.com/loch-ness

*Scanning QR codes requires a web-enabled smart device with a QR code reader app and a camera.